OXFORD
UNIVERSITY PRESS

Great Clarendon Street, Oxford OX2 6DP

Oxford University Press is a department of the University of Oxford.
It furthers the University's objective of excellence in research, scholarship,
and education by publishing worldwide in

Oxford New York

Auckland Bangkok Buenos Aires Cape Town Chennai
Dar es Salaam Delhi Hong Kong Istanbul Karachi Kolkata
Kuala Lumpur Madrid Melbourne Mexico City Mumbai Nairobi
São Paulo Shanghai Taipei Tokyo Toronto

Oxford is a registered trade mark of Oxford University Press
in the UK and in certain other countries

Text © Philip Wilkinson 2003

The moral rights of the author have been asserted

Database right Oxford University Press (maker)

First published in 2003

All rights reserved. No part of this publication may be reproduced,
stored in a retrieval system, or transmitted, in any form or by any means,
without the prior permission in writing of Oxford University Press,
or as expressly permitted by law, or under terms agreed with the appropriate
reprographics rights organization. Enquiries concerning reproduction
outside the scope of the above should be sent to the Rights Department,
Oxford University Press, at the address above

You must not circulate this book in any other binding or cover
and you must impose this same condition on any acquirer.

British Library Cataloguing in Publication Data available

ISBN 0–19–910875-7 Hardback
ISBN 0–19–910876-5 Paperback

1 3 5 7 9 10 8 6 4 2

Printed in Italy

Acknowledgements

The publishers would like to thank:
For the Science Museum: Kevin Johnson, Douglas Millard

All photos reproduced in kind permission of the Science and Society Picture Library
with the exception of the following:
Digital Vision; p10tr, 17tr
NASA; p7br, 8-9c, 12r, 14l, 15tl, cr, b, 16bl, 19

OXFORD
in association with

science museum

space

Contents

Stargazing 6

Circling the Earth 8

People in space 10

Going to the Moon 12

Probing space 14

Space plane 16

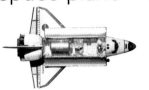

Life in space 18

Space stations 20

Glossary 22

Index 23

Stargazing

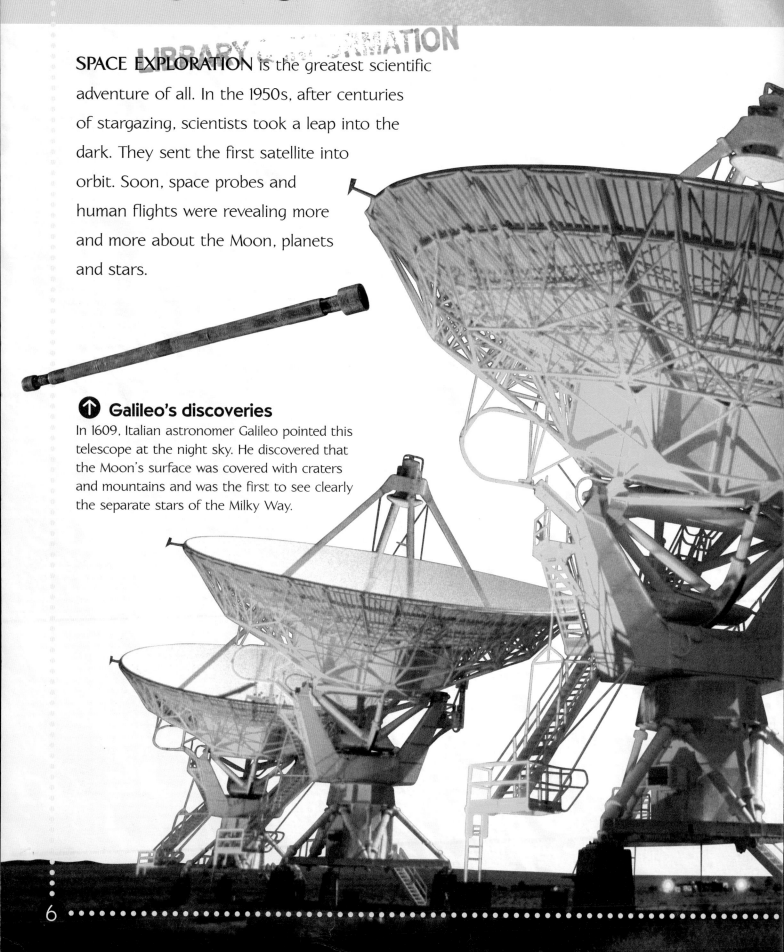

SPACE EXPLORATION is the greatest scientific adventure of all. In the 1950s, after centuries of stargazing, scientists took a leap into the dark. They sent the first satellite into orbit. Soon, space probes and human flights were revealing more and more about the Moon, planets and stars.

⬆ Galileo's discoveries
In 1609, Italian astronomer Galileo pointed this telescope at the night sky. He discovered that the Moon's surface was covered with craters and mountains and was the first to see clearly the separate stars of the Milky Way.

➜ Done by mirrors

Seventeenth-century English scientist Isaac Newton designed an improved telescope. It contained mirrors, which made the image clearer and free of colour fringes found in earlier telescopes that used lenses.

↘ On a stand

Like Newton's, this 18th-century telescope uses mirrors to produce a clear image, without colour fringes. It is set on a stand with a special mount that allows the user to follow objects in the sky as the Earth rotates.

↙ Dished up

Believe it or not, these huge dishes are telescopes. Instead of detecting light, like an optical telescope, they pick up radio waves coming from stars and other objects in space. They can detect very distant stars and events, such as star explosions, that would not be detectable using an optical telescope.

↑ The Hubble Space Telescope

This telescope orbits the Earth. Because its views of the stars or planets are not distorted by the Earth's atmosphere, it is more useful than ground-based telescopes.

Circling the Earth

THE FIRST man-made objects to orbit the Earth were satellites, objects that are launched into space and circle our planet continuously. Since the Russians launched the first satellite, Sputnik 1, in 1957, satellites have been used in all sorts of ways – for example, in weather forecasting, navigation, and sending telephone signals.

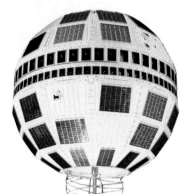

← Telstar
Launched in 1962, Telstar was one of the first satellites that could both send and receive signals. It was used to send TV signals around the world.

↓ High over New York
Landsat took this image of New York City to show land use. Built-up areas appear blue-green, open spaces are red, and water is black. Manhattan is in the middle of the picture and Central Park appears as a red strip in the heart of Manhattan.

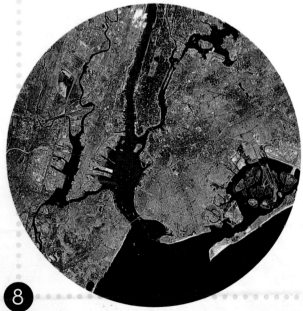

↑ Looking down
Satellites such as Landsat-5 are used to study our planet. Landsat carries special sensors – devices that can detect various kinds of radiation coming from the globe. Computers turn this information into pictures. Scientists can use the results to make maps and to study land use, pollution, the health of crops and forests, and even the rocks and minerals underneath the Earth's surface.

➔ Tracker
This small satellite was launched by the Space Shuttle. As its orbit crosses the Atlantic Ocean, it tracks the movements of whales off the coast of Argentina.

People in space

IN 1961 THE RUSSIANS amazed the world by sending a man into space for the first time. Soon the Americans and Russians were racing to be the most advanced space explorers. They built more powerful rockets, designed better equipment for their astronauts, and learned more and more about how to survive in space.

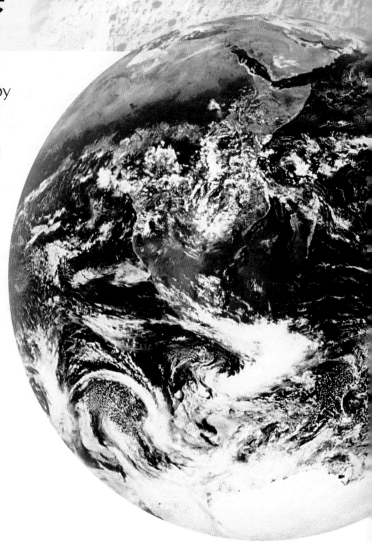

➡ Our planet from space
Space travel gave us a totally new view of the Earth. When people saw photographs taken by astronauts, they were impressed by the beauty of the planet, with its blue seas and swirling white clouds.

↘ First in space
Yuri Gagarin made the first space flight on 12 April 1961. He made one orbit of the Earth before the automatic controls of his Vostok spacecraft brought him back down. His safe return was a triumph for Russia.

↘ Vostok
Gagarin sat in the round capsule at the front of his Vostok spacecraft. Equipment, batteries and Gagarin's air supply were stored in the rear. Only the front section returned to Earth at the end of the flight.

↑ Angry alligator
One skill that astronauts have to practise is docking – linking two craft together in space. This technique can be used in many different situations – for example, when arriving at a space station or when rescuing astronauts from a damaged space capsule. In 1966, US astronauts took this photograph of a craft nicknamed the 'Angry Alligator', with which they were hoping to dock. The astronauts flew their Gemini 9 spacecraft very near the 'Alligator', but its jaws would not open properly, so they could not dock.

← Blast-off
The US crewed space programme began in May 1961, when this rocket launched astronaut Alan Shepard's Mercury craft on its short flight. It went 187 kilometres (116 miles) into space and came straight back down.

Going to the Moon

THE MOON is the Earth's nearest neighbour – but it is still around 384,000 kilometres (238,000 miles) away! Sending an astronaut there was one of the biggest space challenges, but the Americans managed it by 1969. It took a huge team of scientists and several trial runs in the famous Apollo spacecraft before they were ready to make a landing on the Moon's surface.

➡ Surveyor
A safe way to study the Moon was to send an uncrewed, remote-controlled lander, like *Surveyor*. *Surveyor* beamed thousands of pictures to Earth and tested the lunar soil.

↙ We have lift-off
The Saturn V rocket had to be powerful enough to carry three astronauts, their Apollo spacecraft, and their equipment at a dizzying speed of 11 kilometres (6.8 miles) per second. This is the speed at which a rocket can escape the strong downward pull of the Earth's gravity.

➡ First on the Moon
Announcing that 'The Eagle has landed', American astronaut Neil Armstrong piloted his lunar module, *Eagle*, safely down to the Moon's surface in July 1969. Soon afterwards he photographed his colleague, Edwin ('Buzz') Aldrin, as the two collected rock samples and explored the lunar landscape. They spent 22 hours on the Moon before their craft took off again safely.

⬆ Waiting patiently

While Armstrong and Aldrin landed, the third member of the mission, Michael Collins, stayed in the command module, orbiting the Moon. He waited for *Eagle* to return and dock with his craft so that they could all return safely to Earth.

⬆ Moon buggy

Later Moon missions carried a battery-powered vehicle, called the lunar rover. This vehicle could move at up to 14 kilometres (8.7 miles) an hour and enabled the astronauts to explore more of the Moon's surface.

Probing space

SPACE PROBES are the robots of space exploration. They travel far through the Solar System, sending back information about planets, moons and comets that are too distant for humans to reach. Some take pictures as they fly past their target planets. Others are landers that have given us close-up views of Venus and Mars.

⌲ Space pioneer

In 1973, Pioneer 10 was the first probe to fly past Jupiter. Powered by two generators on the ends of its long arms, the probe took hundreds of pictures of the planet and some of its moons.

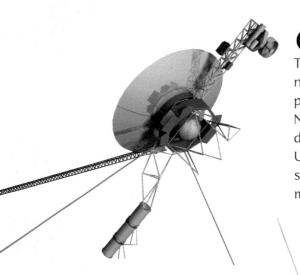

← Long-distance traveller

The two Voyager probes were well named. They voyaged past four planets – Jupiter, Saturn, Uranus and Neptune – during the 1980s. They discovered tiny moons orbiting Uranus and sent back pictures showing how Saturn's rings are made up of hundreds of ringlets.

← The red spot of Jupiter

Pioneer 10 took this stunning picture of the huge red spot on Jupiter. Several times bigger than the Earth, this spot is actually a storm in the planet's atmosphere, which is made of hydrogen and helium gas. The red spot is always on the move, making a full anti-clockwise turn every six days. The probe also sent back information about the powerful radiation given out by Jupiter and the planet's strong, pulsating magnetic field.

↑ Sojourner

The Pathfinder probe landed this vehicle on the surface of Mars in 1997. It sent back information about the red, rocky ground (shown below), discovering magnetic dust and showing that water had once flowed on the planet.

Space plane

THE FIRST SPACECRAFT were used once, then thrown away or abandoned. The Space Shuttle is a craft that can be reused like an aeroplane, saving money and resources. It has a huge payload (cargo) bay to ferry satellites and space station parts into orbit. It is protected by special tiles so that it does not burn up in the scorching heat as it re-enters the Earth's atmosphere.

In orbit
This Shuttle is approaching the International Space Station. It is delivering some crew members and cargo, and the payload bay doors are already open so that the craft can dock with the space station. The Shuttle can stay in orbit for at least 10 days before returning to the ground.

Rocket power
At lift-off the Shuttle is made up of four parts – the orbiter craft, a huge fuel tank, and two rocket boosters. Using fuel from the tank, the Shuttle's main engines and the boosters power the spacecraft into space. The boosters then parachute back to Earth and the tank is discarded.

➡️ Space walk

Astronauts sometimes have to take space walks outside the Shuttle, in order to do jobs such as collecting a satellite for repair. Tasks like this are very dangerous, so astronauts wear a special spacesuit with its own life-support system. Sometimes they work securely tethered to the Shuttle. At other times they wear a special manned manoeuvring unit (MMU), a power system fitted in a backpack, which allows them to move around easily in space.

⬅️ Touch-down

A parachute slows the Shuttle as it lands back on Earth after a journey of almost 6 million kilometres (3.7 million miles). The speeding craft needs 4.5 kilometres (2.8 miles) of runway to slow down.

Life in space

LIKE EVERYONE ELSE, astronauts have to eat, drink, sleep, breathe, move, and go to the toilet. But in space there is no air and no gravity, so you feel weightless. Most of the equipment used by space travellers is designed to cope with these problems.

◆ Space meal
In the weightless conditions of space, everything floats around – even people! These astronauts on the International Space Station are about to share a meal. Their food is wrapped in special containers to stop it moving around the place. Some of the food is dried, so water has to be added, again from a special container. Everything, including the knives and forks, has to be tied or stuck down when not being held by the astronauts.

◆ Instant drink
This pineapple and grapefruit drink was made for the Apollo missions. The drink was dried to save weight. To use it, the astronaut added water and drank the mixture through the tube at the top.

◆ Spacesuit
A spacesuit is designed to keep the user's body heat constant in the extreme temperatures of space. It also has to be connected to a life-support system that supplies the wearer with air.

⬆ Power shower
The Skylab space station had a folding shower unit that was pulled up from the floor. The shower blasted water at the astronaut before sucking it up again.

⬇ Hairy problem
Crew members can spend months on a space station, so they cannot always wait until getting back to Earth for a haircut. This astronaut holds a vacuum device to suck up floating hairs as his colleague snips away.

Space stations

ORBITING THE EARTH for years on end, space stations are like floating laboratories. A space station contains accommodation for the crew, somewhere for them to do scientific experiments, and a place for visiting spacecraft to dock when delivering astronauts or supplies. The latest and largest of these orbiters is the International Space Station (ISS), where men and women from all over the world carry out research in all sorts of areas from astronomy to the effects of weightlessness.

⬇ Sun power

Huge solar arrays make the ISS the size of a football pitch. The seven crew members work in several laboratories, researching the effects of low gravity and developing new technologies.

→ Under construction

The ISS was put together in space using techniques learned in test sessions like this one. Component parts were taken into orbit on the Shuttle. The Shuttle's cargo doors were opened and spacewalking astronauts began to join the pieces together. In 2000, two years after construction had begun, the ISS was ready for its first three crew members. By this time, the space station's basic framework was in position, making it easier to add further parts and build the ISS up to its full size.

Mir

Built by the Russians, Mir was a long-lasting space station that went into orbit in 1987. For more than 10 years, it was used for researching weightless conditions and observing the Earth.

Docking

A visiting spacecraft has to attach itself to one of the space station's docking ports in order to transfer goods and people inside. Astronauts have to fly with great skill to complete this difficult manoeuvre.

Glossary

astronaut Person who travels in space

capsule A small spacecraft, especially the part of a larger craft that contains crew and instruments and can fly as a separate unit

colour fringes Patterns of colour around the image of an object in an optical telescope caused by the way light is bent when it passes through a lens

command module In the US Apollo space programme, the part of the spacecraft in which the astronauts lived during the journeys to and from the Moon and in which they returned to Earth

docking Manoeuvre in which one spacecraft joins up with another, in order to transfer cargo or crew between the two craft

galaxy A large body made up of billions of stars together with gas and dust, held together by gravity

generator Device used to produce electricity

gravity Force that attracts objects to each other; the strength of this force depends on the mass of each object

lander The part of a spacecraft that lands on the surface of the Moon or other heavenly body

Landsat Type of satellite used to gather information about the surface of the Earth

lunar module In the US Apollo space programme, the part of the spacecraft that landed on the Moon

Milky Way The galaxy in which our solar system is found

mount Device on which a telescope is fixed; the mount keeps the telescope steady, while at the same time allowing it to be aimed accurately

orbit Path of one object through space around another more massive object such as a planet or star

radiation Process by which energy, such as heat, light and radioactivity, travels through space

satellite Any object that orbits a planet, but especially a man-made device sent into orbit around the Earth

sensor Device that detects changes in the environment and turns them into a signal that can be measured or recorded

solar cell [or solar array] Device used to collect energy from the Sun and turn it into electricity

solar system The Sun together with the bodies that surround it – the Earth, the other planets, moons, asteroids and comets

space probe Uncrewed, robotic spacecraft, launched by rocket and sent to a specific target, which it investigates using scientific instruments

Index

Page numbers in *italic* type refer to illustrations.

A
Aldrin, Edwin ('Buzz') 12
Apollo 12–13, *12*, *13*
Armstrong, Neil 12
astronauts 10–11, *10*, 17–19, *17*, *18*, *19*, 20, *21*

C
Collins, Michael 13, *13*
communications satellites 8, *8*

D
docking 11, *11*, 20, 21, *21*

E
Earth 10, *10*
  atmosphere 7, 16
  rotation 7

G
Gagarin, Yuri 10, *10*
Galileo 6
gravity 12, 18

H
Hubble Space Telescope 7, *7*

I
International Space Station 16, 18, *19*, 20–21, *20*, *21*

J
Jupiter *14*, 15

L
Landsat 8, *8–9*
lunar rover 13, *13*

M
map making 8
Mars 14, 15, *15*
Milky Way 6
Mir space station 21, *21*
MMU (manned manoeuvring unit) 17
Moon landings 12–13, *12*, *13*
Moon's surface 6, 12

N
navigation 8
Neptune 15
Newton, Isaac 7

O
optical telescopes 6–7, *6*, *7*

P
Pathfinder probe 15, *15*
Pioneer probes 14–15, *14*

R
radiation 8, 15
radio telescopes 6–7, *7*
rockets 10, 11, *11*, 12, *12*, 16, *16*

S
satellites 6, 8–9, *8*, *9*, 16
Saturn 15
Shepard, Alan 11
Skylab 19, *19*
Sojourner 15
solar power 20, *20*
Solar System 14
space probes 14–15, *14*, *15*
Space Shuttle 9, 16–17, *16–17*, 21
space stations 16, 18, *19*, 20–21, *20*, *21*
spacesuits 13, 17, *17*, 18, *18*
space walks 17, *17*
Sputnik 8
stars 6
Surveyor lander 12, *12*

T
telephone signals 8
telescopes 6–7, *6*, *7*
television signals 8
Telstar 8, *8*

U
Uranus 15

V
Venus 14
Vostok 10, *10*
Voyager probes 15, *15*

W
weather forecasting 8